THE LEOPARD
IN CANADIAN SERVICE

Michael R McNorgan

Service Publications
Box 33071 Ottawa, Ontario,
Canada K2C 3Y9
www.servicepub.com

This edition first published in Canada,
December 2005 by
Service Publications,
PO Box 33071
Ottawa, Ontario K2C 3Y9
Ph - 613-820-7350
Fx - 613-820-1288
sales@servicepub.com
www.servicepub.com

ISBN 1-894581-31-8

Printed and bound in Canada
Cover and book design by Clive M. Law

Series editor-in-chief - Steve Guthrie

*Cover photo; A Leopard C2 tank from The Royal Canadian Dragoons, 17 April
2003, Wainwright, Alberta. Photo courtesy Department of National Defence,
LF2003-0658*

Acknowledgements
Thanks are due to the Canadian Forces Joint Imagery Centre,
especially Janet Lacroix and Dennis Mah and to the Directorate of
History and Heritage for their usual unfailing professionalism and
courtesy.

*Michael R. McNorgan served for 39 years in the Royal Canadian
Armoured Corps, in both regular and reserve regiments, retiring as a
major. He is the author of* The Gallant Hussars: A History of the First
Hussars Regiment 1856-2004, *(The Regiment 2004). He is also co-
author, with John Marteinson, of* The Royal Canadian Armoured Corps:
An Illustrated History *(Robin Brass Studio 2000), and a contributor to*
Fighting for Canada *(Robin Brass Studio 2000) and* More Fighting for
Canada *(Robin Brass Studio 2004). He lives in Ottawa.*

A Leopard IA-2 in a German field. One of the early 'rent-a-tanks' that carried the armoured corps through the interim period of January 1977 - October 1978 while the Leopard C1s were under construction. Department of National Defence, Canadian Forces Image Centre (CFIC) GNC00-0134-37

A New Tank

In the 1960s Canada had purchased 274 Centurion Mark III gun tanks, which were eventually upgraded to Mark XI status, as well as nine armoured recovery vehicles and four bridge layers. These vehicles had equipped the regular regiments and the Royal Canadian Armoured Corps (School) in Canada, and the single armoured regiment deployed with the NATO forces in Europe. The Army had estimated that these vehicles would be obsolete by 1971 (*see "The Centurion in Canadian Service"*).

In 1970, a review of the nation's foreign policy led to Canada's ground commitment in Germany - 4 Canadian Mechanized Brigade Group (4 CMBG) - being cut in half, an action that made absolutely no military sense but retained a low-cost Canadian presence. The armoured regiment was now configured with just two tank squadrons instead of the normal three, while the infantry were reduced from three battalions to two with each battalion containing three rifle companies in place of four. The remnants of 4 CMBG were moved 300 miles southwest to the beautiful Black Forest region where they were co-located with Canada's Air Force contingent, at Lahr and Baden-Soellingen.

This action was followed by the 1971 Defence White Paper, which called for the replacement of the Centurions with an air-portable "light, tracked, direct-fire support vehicle." The air-portable replacement that the White Paper had in mind was the British Scorpion, a tracked reconnaissance vehicle with a three-man crew and a 76mm gun. But Scorpion was not a main battle tank, and the Chief of Defence Staff (CDS) was opposed to its acquisition; he wanted a tank. In fact, talks had quietly been opened with the German tank manufacturer, Krauss-Maffei of Munich, to investigate buying their Leopard. The CDS eventually succeeded in stopping the Scorpion acquisition but there was no Cabinet approval for the acquisition of a new tank, at least not yet.

In the fall of 1972, when it appeared the Scorpion deal was dying, Canada secretly reopened negotiations with Krauss-Maffei, and a bargain was struck. *"We could have bought at a guaranteed delivery price"* said the officer in charge, Brigadier-General George Bell. *"160 Mark Is with first year spares for about $320,000 a copy...and had them delivered within two years. By the Fall of 1974 we would have been converted."* Unfortunately, the deal fell through because Cabinet would not authorize the $55M required.

Leopard IA-2 'rent-a-tank'. This crewman is sitting on top of the box that houses the tank searchlight. On the C1 models this position would hold the Low Light Level TV Image Intensifier. CFIC IL77-833

That is where matters stood until the government affirmed, in its 1974 Defence Review Structure, that Canada would remain in NATO's Central Region, the government's earlier actions having cast considerable doubt on that matter. It was then that the West German government began applying pressure to ensure that the Canadian military contribution would be a meaningful one. The West Germans mounted what came to be called the 'no tanks - no trade' doctrine. The Prime Minister, P.E. Trudeau, was vulnerable to this pressure because he was advocating stronger trade ties with Europe as a counter-balance to trade with the US. As a result of these events the Canadian military finally gained both permission and money to purchase a new Main Battle Tank (MBT). Unfortunately, for Canada, this long-sought goal was only reached at an interval in time when the major western tank manufacturers were all in the process of changing over from one generation of MBT to another. The then current generation of tanks, which included the Chieftain, AMX-30, M 60 and Leopard I represented 1960s technology, while the forthcoming vehicles; Challenger, Leopard II and M-1 Abrams, would embody the emerging technology of the 1980s. Should the Canadians buy the available, but soon to be obsolescent, vehicles or wait for the next generation? The problem with waiting was that the government might just change its mind!

The CDS, General J.A. Dextraze, concluded that he should move immediately and, on 20 May 1976, an agreement to buy Leopard IA4s was signed. These Canadian vehicles would be the last Leopard Is produced. The order was for 114 gun tanks, eight Taurus Armoured Recovery Vehicles (ARV) and six Beaver Armoured Vehicle Launched Bridges (AVLB). Even so, there would be a lag in time before these weapons could be placed into Canadian hands. The Centurions were now hopelessly obsolete so Dextraze arranged to lease two squadron's-worth of Leopard IA2s to equip the Royal Canadian Dragoons, the unit then stationed in Germany, through the interim period of January 1977 - October 1978. Of course, after the delays, the price had gone up. These 128 vehicles cost the Canadian taxpayer $187M. By contrast, the deal that had been passed up four years earlier would have seen Canada with 160 tanks at a cost of $55M.

As far as Dextraze was concerned the Leopard programme was merely a stopgap solution to tide the army over until a modern MBT of the likes of the on-coming Leopard II or the XM-1 Abrams (his personal choice) could be secured. This may be why only enough vehicles to

A rear view of the new Canadian Leopard C1. Photo courtesy Phil Waterman

cover Canada's commitments in Germany were purchased. The remainder of the 'land element', as the Army was now called, remained equipped with an ungainly mixture of obsolete (wheeled) Ferret scout cars and the relatively new (tracked) Lynx tracked reconnaissance vehicles; two vehicles whose principal military attribute was their air-portability.

The regular armoured regiments in Canada, and a number of Militia armoured units, would be equipped with the Mowag tank trainer, later christened the Cougar. This was an 11.6-ton, 6x6 wheeled vehicle equipped with a 76mm L23A1 gun. As with the Leopard, the Cougar buy was intended to be a stopgap measure on the way to re-equipping the entire armoured corps with new main battle tanks. But, like the Leopard buy, this scenario was not played out and the Cougar tank trainer actually ended up being used operationally.

The final distribution of Leopards included 57 gun tanks and three Taurus ARVs to the regiment in Germany. Nineteen gun tanks and 1 ARV to the independent tank squadron in Gagetown, 9 MBTs to the School, 6 Beaver AVLBs to the field engineers in Canada and Germany (3 more were purchased in 1989) and eight ARVs to 4 Brigade's Service Battalion. The remaining vehicles were sent to the Canadian Forces School of Automotive Engineering in Borden for the training of technicians and to operational logistics stocks. In 1990 six Badger Armoured Engineer Vehicles (AEV) were acquired.

The Krauss-Maffei AG Leopard

The Krauss-Maffei AG Leopard was the first German tank produced since the Second World War, the prototype being developed between 1959 and 1965. The original model weighed in at 40 tons and mounted an unstabilized 105mm gun that moved through a range of 20º in elevation to 9º in depression. The gunner utilized a stereoscopic range finder. The diesel engine, coupled with a four-speed automatic transmission, allowed a maximum speed of 65kph on roads. The vehicle was capable of deep fording with some crew preparation. The turret was also equipped with an over-pressure system to keep out nuclear, biological and chemical (NBC) contaminants. This meant that it was not necessary for the crew to wear NBC masks as long as they remained closed down inside the vehicle. The Leopard was therefore

A Leopard C1, in winter camouflage, on exercise somewhere in Germany. CFIC IL79-171

capable of passing through an NBC contaminated area.

The next model, the Leopard I A1, saw the introduction of a stabilization system for the main armament permitting engagements on the move, while a new thermal jacket on the 105mm barrel prevented barrel droop and distortion. It also incorporated an item that became a Leopard trademark, the distinctive protective side skirts with foot inserts that the more historically minded tankers referred to as stirrups.

The A2 saw a new, stronger cast steel turret, improved NBC protection, an externally mounted searchlight and the addition of image intensification periscopes for both driver and commander, offering a significant improvement in night operations.

The A3 brought in a new turret, welded this time and a different, more box-like, shape. The new turret added 1.5 cubic metres of much-needed space but most of the room gained was used up with stowage bins for the searchlight, which was now placed under armour, the camouflage net and the gun brushes.

The A4, with some minor modifications, was the version purchased by Canada, had the same basic turret as the A3 but included an Integrated Fire Control System. This system incorporated a fire control computer, stabilized day/night commander's sight and laser range-finder. (There would be a Leopard 1A5 but it was not a true production model. The A5 was actually a hybrid consisting of a new turret mated with an old hull.)

Canada termed the Leopard 1A4 the Leopard C1. It weighed 47 tons and had a range of 490 kilometres (325 miles). The tank's crew consisted of a commander, loader/operator, gunner and driver. For armament it carried a 105mm L7A3 main gun with an effective range of 2000 plus metres and two 7.62mm machine-guns. The Canadian tanks also mounted the PZB 200 low-light level TV image intensification system that allowed for the engagement of targets out to 1200 metres. In addition it carried an ANVSS 4 searchlight mounted inside the turret and a Belgian SABCA fire control system which included a ballistic computer and laser range-finder.

The view from above. Note the camouflage scheme, which stands out well in this photograph. CFIC LPC87-493

The searchlight proved to be a contentious item. Earlier marks of the Leopard had the searchlight mounted externally, whereas the C1 had it placed under armour with a shutter that could close quickly to prevent any after-glow. The established tactics of the day called for the use of searchlights in defensive positions. A designated vehicle would illuminate the target with white light - for 15 seconds - while flanking vehicles would do the firing. All vehicles would then jockey into new positions. However, the growing sophistication of night-viewing devices made the use of white light a very problematical matter against an enemy armed with modern equipment. Although firing with the searchlight was great fun and an interesting type of gunnery training, its best use was likely reading distant German road signs at night!

Likewise, the second machine-gun, which was mounted on top of the turret for anti-aircraft protection, could be a problem. If it was left mounted during road moves it got dirty quickly and, at night, it invariably got entangled in the camouflage net. Most crews kept it stowed, bringing it out only for use on the gunnery ranges.

As mentioned above the first unit to receive Leopards was the Royal Canadian Dragoons, then based in Lahr, Germany as part of 4 CMBG. Their new Leopard I A2s - there would be 35 of them - were nicknamed 'rent-a-tanks' and these started to appear in Lahr on 6 January 1977. In preparation for their arrival, selected officers and non-commissioned officers (both Dragoons and personnel from 4 Service Battalion) had been despatched to German facilities to learn the ways of the new vehicles. These people then had to prepare training packages for the instruction of the Canadian units. This involved much more than just the translation of German training manuals because the professional, volunteer Canadian soldier was taught matters in far greater depth than West Germany's short-service conscripts were. A Dragoon wrote;

"When a trooper climbs up and enters the Leopard turret he enters a world completely different than that of a Centurion. The turret is much more cramped, with barely enough room for the crew and their equipment. Gone were the days when extra rations, camp cots, spare kit and 'a kitchen sink' were the normal companions of a Canadian tank crew.

The gunner has perhaps the tightest squeeze of all. He sits at his futuristic station

Training Tank - Instructor's Cabin. These specially built turrets enabled the School's Driving and Maintenance Squadron to train drivers under operational conditions. The cab holds one instructor and two additional students. CFIC GNC95-420-12

surrounded by two panels of lights and switches. Instead of one control for power traverse and one for power elevation, these elements are combined in one small device that looks as if it was pulled straight out of a jet fighter. The lower turret of the Leopard also allows less space for the loader/operator. In addition, his space is reduced by 11 rounds in the turret's ready racks (as opposed to the Centurion's six).

The world of the driver has changed most dramatically. Although the driver also has less room to move about than he did in a Centurion, there are other benefits. His left foot is freed from the chore of providing 25 kilograms (60lbs) of pressure to disengage the Centurion's clutch.

At the same time, he no longer needs to pull tiller bars to steer. A control column similar to a drag racer's replaces them, with hydraulic assistance.

When the crew commander jumps into his seat and roars 'action', his is a changed world. Not only are the responses of the crew different, but his station has changed as well. It's a tight squeeze, surrounded by light and switch panels, and an armrest for his override control. He has a telescopic sight, which zooms to 20x magnification simply by changing pressure on a button near his right toe. With this sight, the crew commander can search out a new target while the gunner is firing, establish the range to it, then bring the gun around and lay it onto the target and fire. This ability to fire from the crew commander's station is especially important at night, because the crew commander is the only one with an infrared sight."

The new night vision devices quickly proved their worth as Major Ray Richards discovered on one of the first exercises with the new vehicles.

"We were working with a VanDoo (Royal 22e Regiment) company, and they were leading that night, as per doctrine. We were doing a move, and [using our low light television viewers] we started calling out contacts 1500 to 2000 metres away. But the infantry section out front could only see 100 metres in front of them, depending on what ambient light conditions there were. [The company commander] came over and said, 'What are you talking about?' I said, 'Over there in that woodline I count four M-60s and six APCs. He then came over and got into the tank, and saw on the viewer himself all of the targets we were calling out. 'OK,' he said,

An Armour School's Driving and Maintenance Squadron vehicle followed by an M113 Armoured Personnel Carrier. While a few students were undergoing hands-on training in the driver's cab, the remainder of the class followed along in an M113 waiting their turn. Note the second tank following the carrier. The training turret carries a dummy gun to familiarize the driver with its size. CFIC GNC96-253-8

'you guys lead at night.' It was outstanding. It was like a shooting gallery along that woodline. We hammered away at what we could see but there was no return fire and you could see these guys diving and jumping around, wondering what was going on.

Then an umpire jumped up and said 'You're all dead.' We replied, 'What do you mean?' They haven't even returned fire yet!' He said, 'That was a reinforced tank battalion, and you've only got four tanks.' I told him it didn't matter how many vehicles they had if they didn't pull their triggers. I showed him the scope, and he thought it was a painted picture! He just couldn't believe it. He then said 'OK, I'll go see how many casualties I can assess.' He came back and said, 'I hate to tell you this but you really are all dead. There were fifteen TOWs (Tube-launched, Optically-tracked, Wire-guided - a wire-guided anti-tank missile) with thermal imaging three kilometres over there on your right. But I've given you credit for all the damage you did back there."

It was in 1977, while the Dragoons were equipped with the 'rent-a-tanks', that their team, commanded by Captain T.J.I. 'Tom' Burnie, won glory capturing NATO's top gunnery prize, the Canadian Army Trophy.

The Leopard C1

The new year of 1978 saw the arrival of the first of the Leopard C1s in Lahr. These tanks were a step ahead of the A2s then being rented in that they incorporated a laser range finder, a ballistic computer which helped lay the gun by taking account of such variables as wind speed, atmospheric pressure, air and powder temperatures, cant angle of the gun (caused when one track is lower than the other because the tank is sitting on uneven ground) and barrel wear. Also attached were a searchlight and a low light-level television sighting system. As each troop of four C1s arrived, the Dragoons tidied up four of the rentals and handed them back to the Germans. By March 1979, the regiment's entire allotment had been received, enough vehicles to mount *three* tank squadrons. As previously mentioned, in 1969 the armoured regiment in Germany had been reduced to just two tank squadrons, each equipped with 19 gun

The turrets removed from the School's Driving and Maintenance Squadron vehicles found their way onto special stands used for training in Gunnery Squadron. Courtesy Phil Waterman

A turret-less hull. They certainly went faster with the turret removed! CFIC PB87-108-2

tanks. Although Dextraze had not been able to secure enough tanks for all of his armoured units he did buy enough to completely equip a three-squadron regiment. The vehicles for this third squadron would be kept in operational readiness in Lahr and maintained by a cadre of 23 personnel. The soldiers ear-marked to crew these tanks in battle were stationed in Canada but were ready to fly-over to Germany, in the event of trouble, to man these vehicles.

One cultural change brought about with the new vehicle was a prohibition on what the crews were allowed and not allowed to do with their vehicles. Formerly, a tank crew on a Centurion was expected to handle virtually any kind of repair, whether it was officially considered a matter for the crew or not. The Royal Canadian Electrical and Mechanical Engineers (RCEME - pronounced reemy) provided second-line support, but the RCEME were

A Leopard C1 going through a vehicle obstacle course. Timing the crossing of the knife-edge in such a manner that neither tank nor crew was damaged took skill and timing. CFIC IL83-158-1

not always around and Centurions would break down, in large numbers, in odd locations at any time of the day or night. With the Leopard however, the list of things the crew was expected to leave to the RCEME was an exhaustive one, and sometimes exasperating for soldiers used to handling matters on their own. Crews were continuously reminded that any non-authorized repairs would 'void the warranty.' On the positive side the modularized Leopard was fairly simple to maintain and the entire engine could be changed in just twenty minutes, an activity that had consumed days with the Centurion.

Something new was the painting of a black maple leaf, outlined in white, on both sides of the turret. These were added at the factory. Within the unit, the vehicle's radio call sign would be painted on both sides of the hull, at the rear and occasionally on the back of the turret. Vehicles were also given names by their crews. These names started with the same letter as the squadron's designation and were approved by the squadron commander. Some crews placed the vehicle's name on their mounts; others did not, as the policy varied from unit to unit and time-to-time.

About the only thing Leopard crewmen found that they did not like was the crew commander's seat. In a Centurion this seat was mounted on a slanted bracket. With the hatch open and the seat at maximum elevation, the commander's back rested against the rear portion of the hatch. On lowering the seat the commander's body was moved forward so that his head was placed directly in front of his sight. With Leopard the seat moved straight up and down. This was fine when the hatch was open and the commander's head was outside, but on closing down it meant that in order to look through his sight the commander had to lean forward. Sitting in this position for long hours caused back and neck pain. The Leopard turret was so tight for space, however, that it was not deemed possible to correct this matter.

The Service Battalion, with their Taurus armoured recovery vehicle or ARV, also found themselves in a new world. Although called a recovery vehicle, that was only one aspect of the vehicle's capabilities. Built on a Leopard chassis, the Taurus actually had a much different superstructure. The left hull front was raised, providing protection for the crew of four. On the right hull front, was an extensible hydraulic crane capable of traversing through 270°. When the crane was in use a nose-mounted dozer blade dug into the ground to provide stability. This blade also anchored the vehicle when its main winch (primarily used for pulling other vehicles from ditches) was in use.

Leopard C1

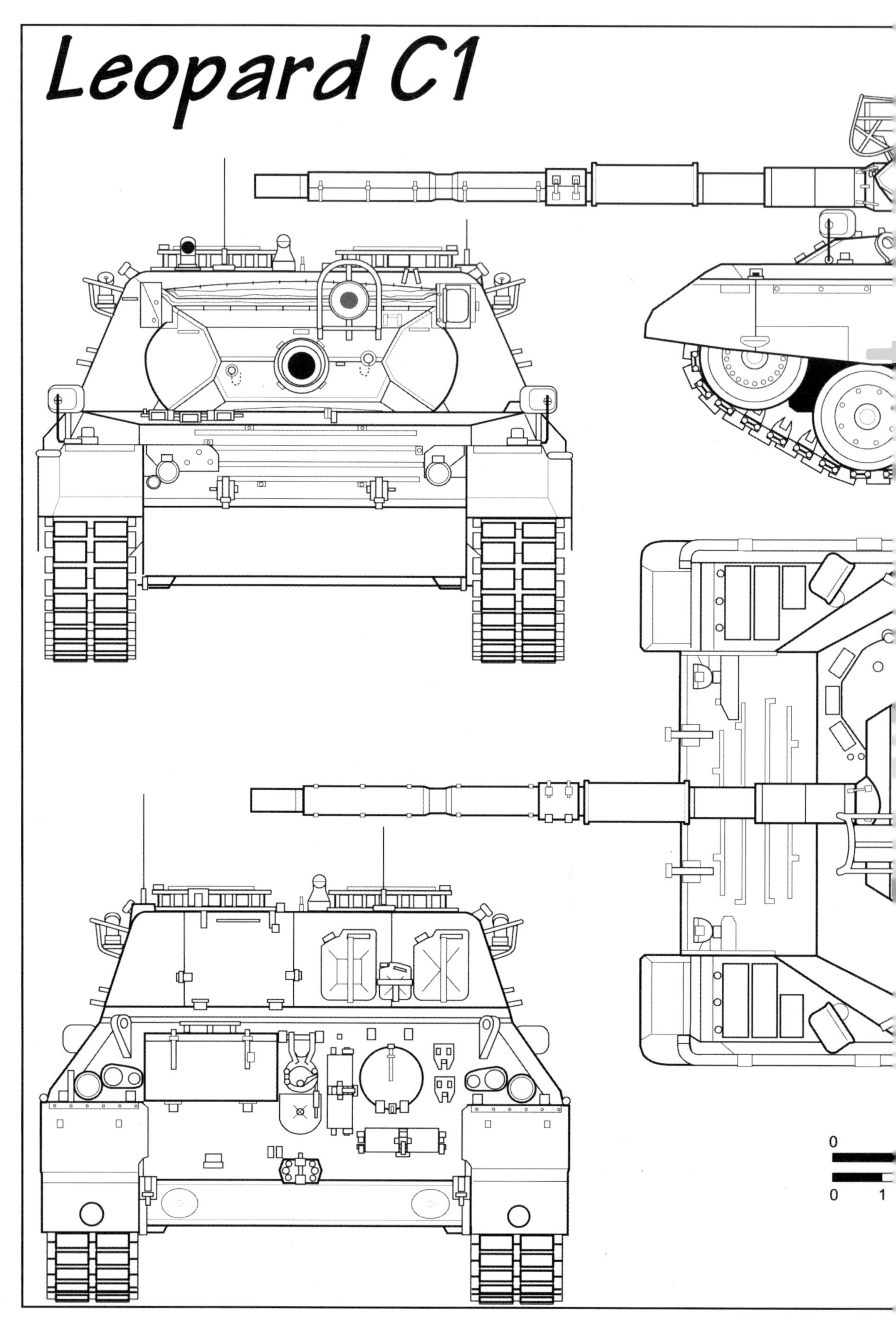

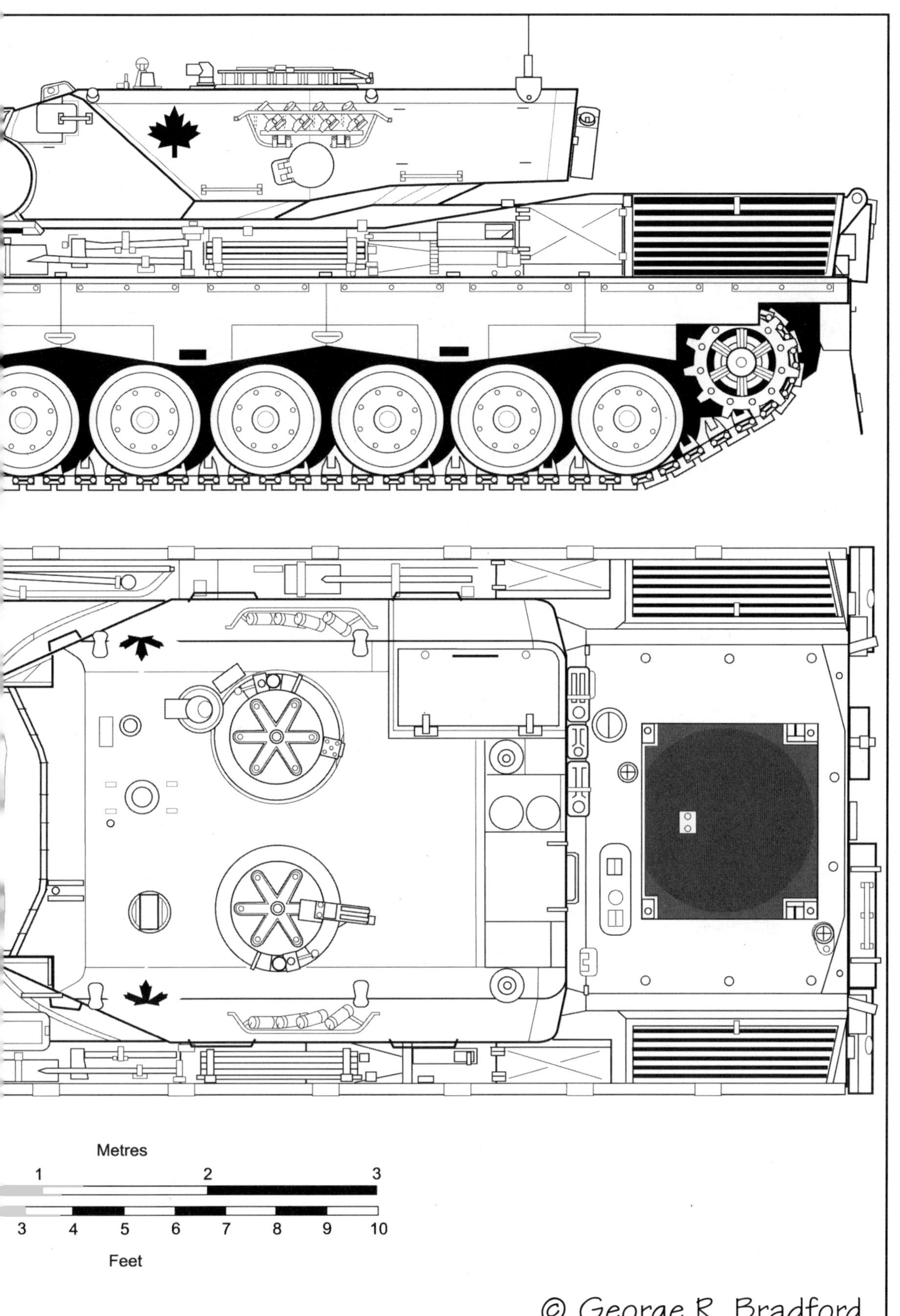

Metres
1
2
3
3
4
5
6
7
8
9
10
Feet
© George R. Bradford

A tank of the Royal Canadian Dragoons during NATO's Canadian Army Trophy competition, widely recognized by NATO tankers as the 'Olympics of tank gunnery'. CFIC IL81-186

Leopards in a flank role with their guns over the side as they move along a road. The shrouded anti-aircraft on the turret top tells us that it is an exercise. CFIC IS81-705

The crane was capable of hoisting up to 20 tons of dead weight; one of its more common tasks being the lifting of complete engines for replacement in the field (a spare engine was commonly carried in a cradle located above the Taurus' engine compartment) The crane could also lift off the turret of a main battle tank! At 40.6 tons, it was the lightest of the Leopard family and had the greatest range - 850km.

The Independent Leopard Squadron

A line of Leopards. Preparing a squadron for an administrative (that is, non-tactical) move. CFIC IL82-208-1

In the summer of 1978 the first personnel allotted to the new independent tank squadron started to arrive in Gagetown. The unit that they were posted to was titled 'B' Squadron, 8th Canadian Hussars and their mandate was two-fold. Operationally they were to train and be prepared to deploy to Germany, on short notice, to man the third tank squadron of the RCD. Their secondary role was to supply vehicles and crews to the Armour School in Gagetown to assist in the running of courses, a function known as being 'school troops.' At first, the new Leopard squadron had no vehicles at all as Germany had the priority for equipment. Therefore, they drew a troop's worth of Centurions from the School in order to get the crews used to working with each other. 'B' Squadron thus became the last Canadian unit to use the Centurion tank as well as the last unit to fire it.

Starting in the autumn of 1978 the Leopard C1s began arriving, one at a time, in Gagetown. The first vehicle to appear was a subject of considerable awe and much conversation with everyone itching to get inside to see what it was like. The Minister of National Defence, the Honourable Barney Danson, came to the ceremonial parade, on 16 October 1978, mounted by the School and 'B' Squadron to mark the momentous occasion. As Danson, who had lost an eye serving in Normandy with the Queen's Own Rifles, stood on the reviewing stand the last of the venerable Centurions creaked and groaned their way past, dipping their guns in salute. Once safely by the reviewing stand and out of sight around the corner of the tank hangar one of them finally gave up the ghost, a wispy column of smoke marking the place where it had died. The single new, recently arrived, Leopard now roared around the corner into view, its inexperienced crew, driving too fast, made a pivot turn at the last moment to proceed by the reviewing stand. The gun traversed in salute and as it dipped the officers on parade started to see their careers evaporating before their eyes. As the gun came round its muzzle looked to be in line to smack the Minister square in the head. Fortunately for all concerned the muzzle missed the Minister, who did not even flinch, and the parade was safely concluded.

The Hussars worked diligently at both assigned tasks, then, in August 1979, came the first fly-over. Six weeks were spent in Germany during the FALLEX or Fall Exercise, which was NATO's major annual manoeuvre. Working as part of a regimental structure was a new experience for the fly-over troops, but they certainly did not take a back seat to the Lahr-based Dragoons. It quickly became obvious that the fly-over squadron's tactical skills were of a

15

An ARV in Germany. The tank crews' best friend, the Armoured Recovery Vehicle (or 'ARV,' as it was more commonly referred to). CFIC IL77-743

A recovery in progress. The ARV uses its bulldozer blade as a brace in order to lift the gun tank. CFIC ILC79-424

slightly higher level than those of the rest of the unit. This was not because they were better soldiers but due to the fact that they had superior cross-country training facilities. Tactical manoeuvres in Germany were restricted by the need to avoid 'manoeuvre damage', that is damage to crops and civilian property. Any such damage was promptly paid for by Canada but obviously there was an emphasis on causing as little destruction as possible. This inevitably meant minor tactical compromises, such as stopping on a road when one really should be

An Armoured Recovery Vehicle with its tent on the back deck. The tent provides some slight shelter from the elements and is heated by the engine beneath it. CFIC IL79-188

A Badger armoured engineer vehicle with its digging bucket deployed. Note that the bucket is on a retractable boom and that the vehicle also comes equipped with a bulldozer blade. CFIC RVC92-321

positioned in the middle of a farmer's field. The crews from Gagetown did not have these restrictions in their training area and so were better versed in using ground than their Europe-based counterparts.

The fly-over squadron was responsible for the re-introduction of live field firing (as opposed to firing on static ranges) in Gagetown after a hiatus of a number of years. This was the occasion for much nervousness on the part of the staff lest there be any unfortunate accidents. However, the training went off without a hitch and improved the professional standards of the crewmen who carried it out.

The Gagetown tankers soon developed a close working relationship with 22 Field Squadron, Gagetown's engineer unit. The field squadron was equipped with Beaver AVLBs and later the Badger AEV and used both items in field exercises with the tank squadron.

The Beaver's bridge was broken into two sections, which deployed horizontally. A hydraulically driven arm at the front of the vehicle launched the bridge sections while another

Above, An AVLB deploying its bridge. Note the engineers assisting with the task.
Below, The bridge is launched and the tanks will soon be crossing the obstacle.

mast at the rear provided support, a versatile, blade on the front adding stability. Once the two sections locked together, they formed a cantilevered bridge that was 22 metres long and 4 metres wide, capable of allowing a main battle tank to cross a 20 metre-wide obstacle. The crew of two were able to do everything from within the vehicle, the entire operation taking between 2 and 8 minutes to complete.

The Badger AEV was similar to the Taurus. Like the ARV, the Badger's crew was housed in a built-up armoured casement. It also had a hydraulic arm mounted on the right hull front;

An AVLB with its bridge in the stored position. Note the bulldozer blade in front, which served as a brace when deploying the bridge. CFIC RVC92-509

however, its telescoping arm was equipped with a large excavating bucket, the bucket arm being capable of traverse through 180° and up to 60° of elevation. It had a pull-strength of 15 tons and a push-strength of 10 tons - useful in carrying out demolitions. As with the Beaver all work could be done from within the protection of the vehicle.

May of 1980 saw 'B' Squadron, 8th Canadian Hussars re-badged as 'C' Squadron, Royal Canadian Dragoons. At that year's regimental gun camp at Bergen-Hohne, for the first time since the Second World War, the Dragoons went on parade with five full squadrons, three of them mounted on MBTs.

The role of the fly-over squadron ended in 1985 when the government of Brian Mulroney authorized an increase in the manning levels of 4 CMBG in Germany. The RCD Reconnaissance Squadron was re-roled to become 'C' Squadron, taking over that squadron's Leopards. The 8th Canadian Hussars Reconnaissance Squadron was moved en masse to Europe and re-badged to RCD. The 'other' 'C' Squadron, RCD in Gagetown, kept the same name but was now primarily tasked to support the Armour School. In 1985, in place of a flight to Europe, that squadron found itself deployed to Western Canada for Exercise Rendezvous 85. This was a long overdue step toward returning the Canadian military to large-scale combined arms training. For about a year the Dragoons had two sub-units called 'C' Squadron, both mounted on MBTs! Eventually, the Gagetown squadron, which had started life as 'B' Squadron 8th Canadian Hussars, would return to that affiliation but this time as 'A' Squadron.

The Armour School

The School, located in Gagetown, New Brunswick, was issued with nine main battle tanks, which were split between Gunnery and the Driving and Maintenance (D&M) Squadrons. Split in numbers and split literally as the turrets were removed from the D&M tanks so that they could be fitted with a special 'Training Tank - Instructors Cabin', which contained space for an instructor and two additional students. The instructional cabin contained displays that showed the instructor what the student driver was doing and the condition of the vehicle, plus override controls so that the instructor could take over if the student driver got into difficulties.

Those turrets removed from the D&M vehicles were placed on instructional stands in Gunnery Squadron for the training of turret crews. Standing outside the stands instructors could observe and correct the drills of trainee commanders, gunners and loaders.

Officially there was no change in tank tactics after the conversion to the Leopard. However, tactics continued to evolve as crews discovered the potential of this new tank.

A Leopard equipped with a mine roller. Mounted in front of the tank, the rollers would set off any pressure-sensitive mines while the weighted chain between the rollers activated any trip wires strung across the tank's path. CFIC EEC82-1775

Probably the greatest change was in night operations. No longer were tanks virtually blind from dusk to dawn. The idea of armour leading at night instead of infantry has already been mentioned above. On the ranges it was found that the firing of a Sabot (armour piercing) round left a small white dot on the TV screen. If the first round should miss, an increasingly rare phenomenon with this tank, traversing the turret so that the small white dot was placed over the target produced a second-round hit. Opportunities to do realistic training with mechanized infantry and armoured engineers were warmly welcomed but proved to be fleeting because the first priority in Gagetown was the conduct of courses. Similarly, in Germany, the training opportunities were restricted by the small size of the military manoeuvre areas as well as the need to avoid damage to civilian property. Canada actually had large manoeuvre areas at Canadian Forces Base (CFB) Wainwright, Alberta and CFB Petawawa, Ontario but there were no tanks stationed in Canada. Meanwhile our allies, the British and Germans, flew over to Canada every year to train in mechanized warfare in the wide-open spaces of CFB Suffield, Alberta and CFB Shilo, Manitoba!

Prior to 1970, units serving in Europe were rotated periodically back to Canada, a system that ensured that all regiments got an opportunity to experience different missions and locales. In those days, officers and men would serve the majority of their careers wearing a single cap badge. In the other combat arms this was not a problem as all gunners wore a common artillery badge; while all sappers wore a common engineer badge. The multi-battalion infantry regiments were large enough, and insular enough, that once someone joined an infantry regiment they could spend an entire career wearing that regiment's insignia. Armour was somewhat different since armoured units were smaller and each had a distinctive identity. Although it was possible to transfer people between units (it was more common to move officers than soldiers) most preferred to identify with one regiment - a phenomenon known as cap-badge loyalty. After 1970, personnel, but not units, were rotated in and out of Europe. As a result of this policy the units in Europe gradually lost their individual identities and became mere administrative entities, not regiments that commanded their soldiers' loyalty in the traditional sense. After a number of years in Germany the RCD consisted of people who were Strathconas, Hussars or 12ieme Regiment Blinde du Canada (12RBC), temporarily wearing Dragoon insignia, a situation that gave rise to the regiment being referred to, only half

A Leopard C1 crossing a bridge laid by a Beaver Armoured Vehicle Launched Bridge. Note the turret machine gun on the loader's side. CFIC IMC79-192

humorously, as the Royal Lahr Dragoons. This was the background to Operation Springbok/Coronet in 1986, the rotation of the RCD with the 8th Canadian Hussars.

In June 1985, the Reconnaissance Squadron of the 8th Canadian Hussars arrived in Lahr where they quickly found themselves re-badged to become the RCD Reconnaissance Squadron. The personnel from the original Reconnaissance Squadron moved across the North Marguerite to become 'C' Squadron. This resulted in the Dragoons having two squadrons bearing the same name, at the same time. This situation continued until the RCD rotated back to Canada. Not all of the Dragoons went home, however, as approximately 250 officers and crewmen stayed in Lahr, re-badged as 8th Canadian Hussars.

The outbreak of war the Middle East, following the Iraqi invasion of Kuwait in 1990, saw the creation of a United Nations military coalition prepared to take the field, the first since the Korean War. Canada supported the operation and staff checks were made to determine the size and structure of a potential Canadian ground force contribution. Dubbed Operation Broadsword, the plan was built around an armoured regiment (8th Canadian Hussars) and a reconnaissance squadron (Strathconas) along with a three-battalion strong mechanized brigade. Everyone recognized that Leopard was no longer a first-line MBT, and might have to be used in a flanking task or in a rear-area security role. Consideration was also given to placing Canadian crewmen into another vehicle such as Leopard II or Abrams, but in the end it all came to naught. The staff planners, basing their casualty forecasts on data taken from a scenario predicated on conventional warfare against the Soviets in Europe, completely spooked the politicians with dire predictions of mass casualties and the government decided that no Canadian combat troops would go to war. In the event, the 100-hour long campaign, although not without cost, was a virtual walkover with the coalition's armoured formations sweeping all before them. The Canadian land force (as the Army was now called) contribution was a couple of infantry companies sent out as guards for airfields, headquarters and the field hospital.

The collapse of the Soviet Union in 1991 led to the end of Canada's watch on the Rhine and, with the 1992 closeout of Canadian Forces Europe, the last Canadian tank round went

Above, The mine plough with the plough in the raised position. Note the weighted chain for tripping wires and tilt-rod mines.
Below, The mine plough in the down position.

down-range at Bergen-Hohne on 27 November 1992, fired by Lieutenant-Colonel C.S. 'Chuck' Oliviero of the 8th Canadian Hussars. The Leopards were loaded onto ships and returned to North America to close another chapter in Canada's armoured history.

The Post-Europe period

When the tanks came back to Canada, 19 MBTs, that is one squadron's-worth, went to each of the regular regiments, respectively located at CFB Valcartier, CFB Petawawa and CFB Edmonton. In the case of the 8th Canadian Hussars, their sub-unit was stationed at CFB Gagetown as 'A' Squadron, utilizing the vehicles of the fly-over squadron. The Hussars became what was termed a 10/90 unit, composed of 10% regulars and 90% reservists.

The Cold War had ended and Canada was quick to slash military spending as part of their 'peace dividend'. However, the world remained a dangerous place. The end of Communist rule in Yugoslavia led to rising ethnic tensions and an outbreak of fighting. Canada joined in the

A gun tank with a bulldozer blade. One vehicle in squadron headquarters was equipped with a bulldozer blade for the digging of hasty firing positions. CFIC LR90-26-9

international effort to stop the killing, even as the preparations to leave Europe were underway. Canadian soldiers would become quite familiar with the territory of the former Yugoslavia. In the mid-1990s Canadian armoured troops, mounted on Cougar tank trainers, had watched with fascination as a squadron of Danish Leopards destroyed five Bosnian Serb tanks in a series of engagements near Tuzla. Canada had never deployed heavy armour on peacekeeping missions. That policy, however, would change as peacekeeping now all too often became peacemaking and Canadian soldiers found themselves under fire with no direct-fire support capability to back them up.

In July 1999, four Leopards were shipped to Kosovo to provide some armoured backup to the Canadian contingent. First, a Strathcona troop and then a Dragoon troop deployed on Operation Kinetic. As the RCD regimental magazine, Springbok, pointed out;

"The tanks proved themselves time and again. No one expected our old Leopards to withstand the thousands of kilometres of abuse as well as they did. It was rare to have a day when all four tanks were not serviceable and ready for action. This was thanks to our solid maintenance routine, dedicated team of maintainers, and the fact that each tank rolled track nearly every day. Nothing in Kosovo matched the tanks when it came to providing a military presence. Show of force was our middle name. Whenever the tanks went - and they went almost everywhere in the AOR (Area of Responsibility)- the whole grid square knew about it."

The Leopard C2

In the late 1990s the Canadian Leopards were showing their age. The vehicles were twenty years old and their technology well out-of-date. With the end of the Cold War the army had lost its opportunity to upgrade its MBT fleet. The government had slashed the defence budget, yet again, and was certainly not interested in buying a new tank. The downsizing of the German army, however, had left our NATO ally with many surplus Leopard I A5s. The Canadians were not so much interested in the A5s themselves, but in their turrets. (As mentioned above the A5 was a hybrid variant incorporating a redesigned turret with the old A1 hull.) The A5 turret was of a newer ballistic shape and employed spaced armour dubbed MEXAS – 'Modular Expandable Armour System'. Even more important than these welcomed improvements was the EMES18 digital fire-control system that employed a thermal night sight. These were the same systems used in the Leopard II. A Canadian improvement was the addition a spall liner. This liner was designed to better protect the crew from the effects of a

The next generation, a Leopard C2. CFIC HD00-176A

hit on the turret, which might send pieces of the turret wall (termed spall) whirling about the crew space.

Canada paid $145M for the 123 A5s and then swapped their old C1 turrets for the newer ones. The resulting hybrid, dubbed the Leopard C2, was unveiled in 1999 and distributed to the regiments starting in the summer of 2000.

At the same time Canada purchased 18 Leopard Crew Gunnery Trainers from the Netherlands. These permitted the training of commanders and gunners using electronic simulators that could reproduce terrain and weather conditions as well as different operational scenarios. By hooking the simulators together in a local area network it was possible to train a tank troop in basic tactics. The simulators were distributed between the School and the Leopard-equipped regiments.

Postscript

In March 2003 a reorganization of armoured corps assets saw the RCD and 12RBC Leopards shipped west to the Lord Strathcona's Horse, while the RCD switched over to the Coyote reconnaissance vehicle. Changes are in the wind, however, and the LdSH are now part of an experiment to create the direct-fire unit of the future. The Leopard will not necessarily be a part of that future as its service life is nearing its end. The Leopard is expected to be declared obsolete in 2010 or perhaps 2015. Its successor is slated to be the Mobile Gun System or MGS, a light, air-portable, wheeled direct-fire weapon with a 105mm main gun. The MGS proposal will see $600M spent to acquire 66 vehicles.

In August 2005, seven Leopards were sold, at auction, by Crown Assets Disposal Corporation. These tanks were turretless and were sold "as is, where is". Four were in Montreal while the remaining three were at Denwood, Alberta. They averaged less than $15,000 each.